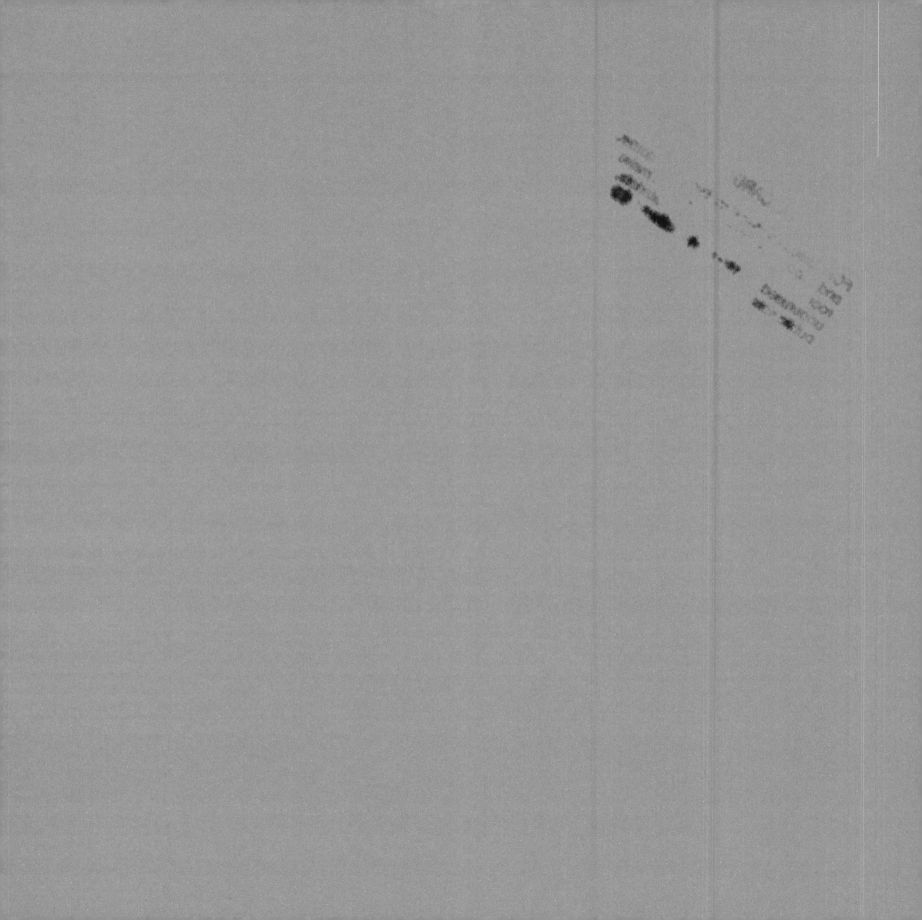

STEAM & Me™

HORSES

BEN GROSSBLATT

Starry Forest Books

SCIENCE · TECHNOLOGY · ENGINEERING · ARTS · MATHEMATICS

Draw a super-smart robot. Create your own wind energy. Find out if your teeth are as sharp as a shark's. Go back in time to the world of dinosaurs or rocket into space. Power up that scientific brain of yours with **STEAM&Me**!

Photos, facts, and fun hands-on activities fill every book. Explore and expand your world with science, technology, engineering, arts, and math.

STEAM&Me is all about you!

Great photos to help you get the picture

New ideas sure to change how you see your world

Let It Snow!
Dapple-gray horses almost look like they're covered in snowflakes.

Palomino

The Pink Horse
Horses with a rose-gray coloring can appear pink.

Blaze

Star

Every horse is beautiful.
Horses' faces are sometimes marked with splashes or flashes of white. Each kind of marking has a special name. The markings are called stars, blazes, snips, and strips. Horses' coats are different colors, and each color has its special name, too. Every horse has its own combination of markings and coat color that make it unique.

STEAM&Me Pick two pictures of horses from this book to compare and contrast. Look at them carefully. What do they have in common? How are they different from one another? Look at their faces, markings, coloring, and size.

18

19

Fascinating facts to fill and thrill your brain

Hands–on activities to spark your imagination

Saddle up!

People all over the world love horses. These animals are strong, fast, and beautiful. It's fun to watch them run and jump. Some people even love watching horses swishing their tails back and forth and lazily nibbling on grass. Did you know that horses and people work together in lots of different ways? That's another reason why horses are so loved.

Giddy-up! In boxes like these throughout the book, you'll find activities and experiments that will help you understand horses better and explore some of the ways you are different from horses—and what you have in common.

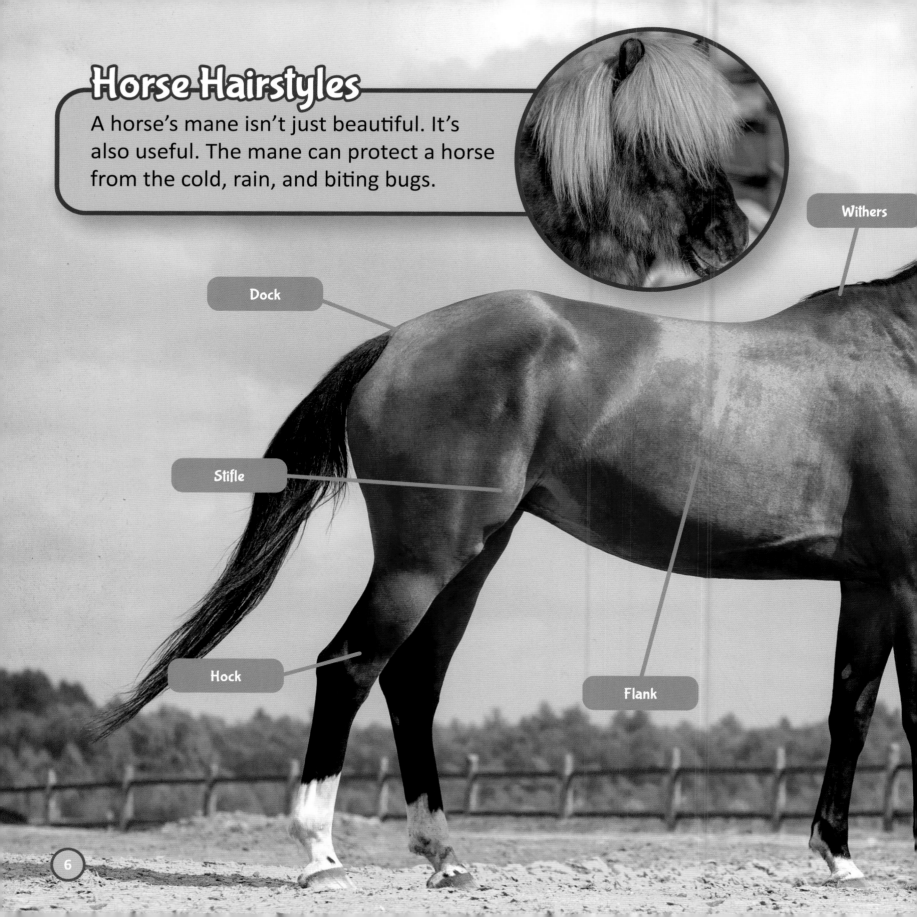

Horse Hairstyles

A horse's mane isn't just beautiful. It's also useful. The mane can protect a horse from the cold, rain, and biting bugs.

Withers

Dock

Stifle

Hock

Flank

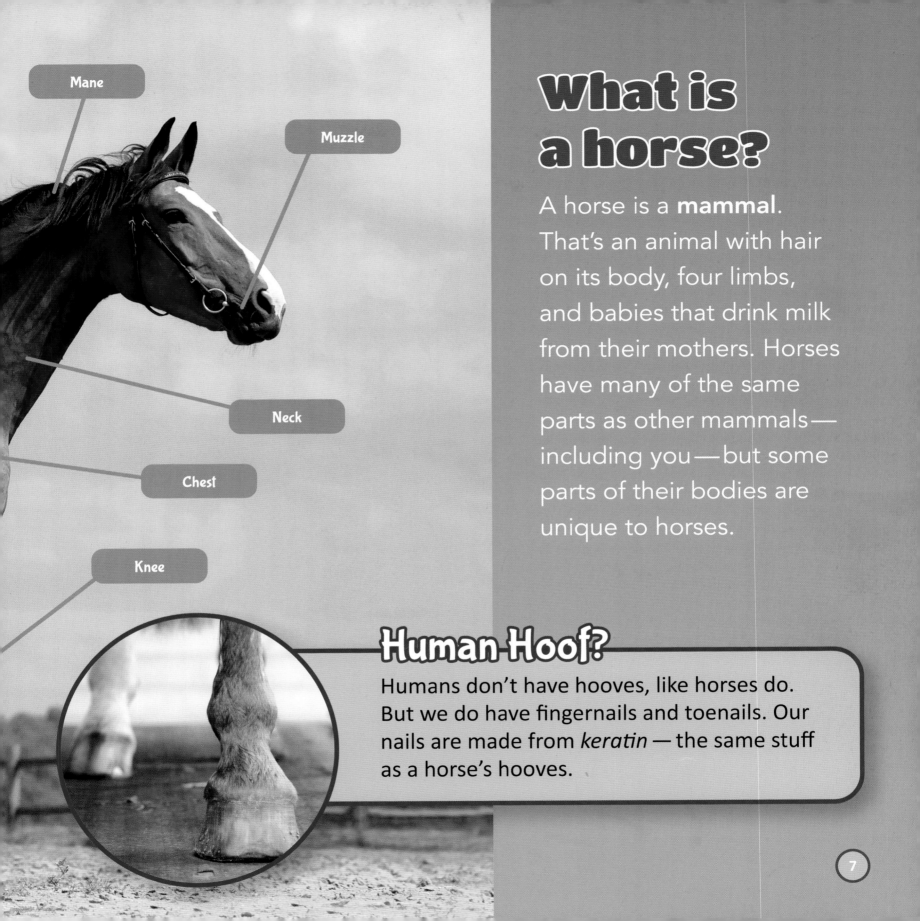

Mane

Muzzle

Neck

Chest

Knee

What is a horse?

A horse is a **mammal**. That's an animal with hair on its body, four limbs, and babies that drink milk from their mothers. Horses have many of the same parts as other mammals— including you—but some parts of their bodies are unique to horses.

Human Hoof?

Humans don't have hooves, like horses do. But we do have fingernails and toenails. Our nails are made from *keratin* — the same stuff as a horse's hooves.

Some horses are BIG and some are small.

Do you know how tall you are? If you do, someone probably measured your height from your feet to the top of your head. But that's not how we measure horses. We measure a horse's height from the bottom of its front feet straight up to the highest part of its back, called the *withers*. That's right: a horse's long neck isn't included. That would be like measuring your height from your feet to your shoulders and ignoring your head!

Horses and Hands

A horse's height is measured with a unit of measurement called a *hand*. A hand is 4 inches — that's about the width of an adult's hand, including the thumb.

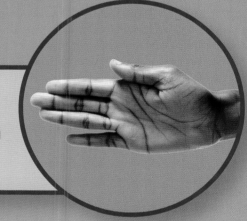

Use a ruler to find out how tall you are from your feet to your shoulders. Have a friend or an adult help you. Are you taller than Hyracotherium? Are you as tall as a modern Falabella horse? Can you figure out how many hands tall you are? Have a grown-up help you.

STEAM *&Me*

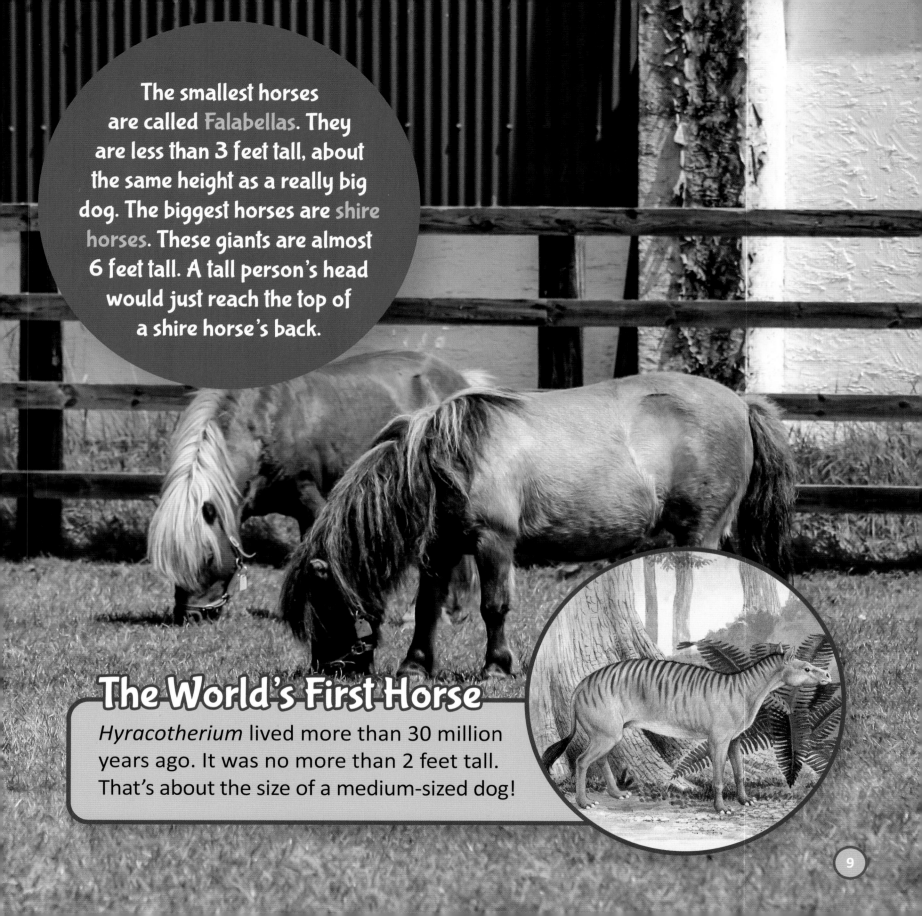

The smallest horses are called Falabellas. They are less than 3 feet tall, about the same height as a really big dog. The biggest horses are shire horses. These giants are almost 6 feet tall. A tall person's head would just reach the top of a shire horse's back.

The World's First Horse

Hyracotherium lived more than 30 million years ago. It was no more than 2 feet tall. That's about the size of a medium-sized dog!

Nighttime Is the Right Time

Foals are often born at night. It takes a few hours for foals to be strong enough to walk, so the darkness might help the mother and newborn to stay hidden and safe while they wait.

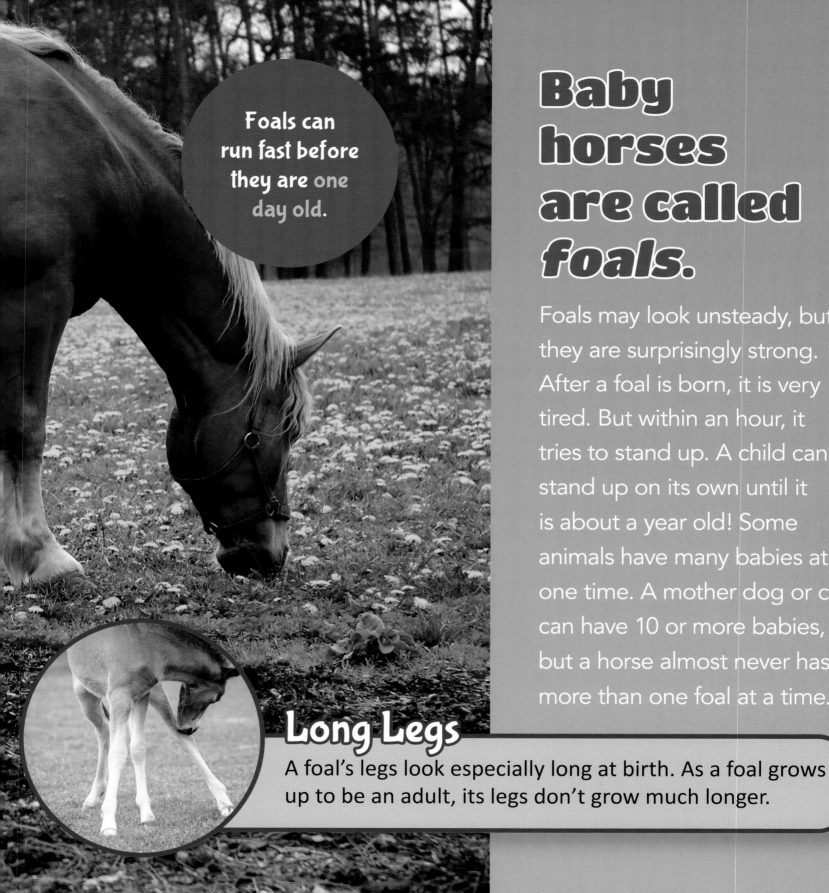

Foals can run fast before they are **one** day old.

Baby horses are called *foals.*

Foals may look unsteady, but they are surprisingly strong. After a foal is born, it is very tired. But within an hour, it tries to stand up. A child can't stand up on its own until it is about a year old! Some animals have many babies at one time. A mother dog or cat can have 10 or more babies, but a horse almost never has more than one foal at a time.

Long Legs
A foal's legs look especially long at birth. As a foal grows up to be an adult, its legs don't grow much longer.

Run like the wind!

Horses are almost like running machines. Their bodies make them great at running fast and far. The fastest racehorse could sprint at about 44 miles per hour. That's almost as fast as a car zooming down the highway.

A horse has a big heart that can weigh 10 pounds or more. A human's heart weighs less than a pound. A horse's big, powerful heart can pump lots of oxygen-filled blood to fuel its muscles.

What happens inside your body when you exercise? Find a place at home where you can safely do 20 jumping jacks. Are you breathing faster now? When your muscles work harder, they need more oxygen. That makes your heart and lungs work harder, too.

STEAM at Me

Winning by a Nose

A horse's big nose and nostrils let it breathe in large amounts of oxygen. More oxygen means more energy for running.

Way to Go

Horses use different **gaits**, or ways of walking and running. A few types of gaits are called *trot*, *canter*, and *gallop*. To run in these different gaits, a horse moves its legs at different rhythms.

Tooth Truth

Veterinarians are animal doctors. They're often called vets. A vet can check a horse's teeth and gums to make a guess about the horse's age and health.

A horse can eat 25 pounds of food in a day. You probably eat about 3 pounds every day.

Come and get it!

Horses are **herbivores**, which means they mostly eat plants. With their wide, flat teeth and strong lips, they can pull up and chew lots of grass and hay. They also eat grains, such as oats and corn.

A Little Treat

Horses like to eat treats just as much as you do. Some of a horse's favorite treats are apples, carrots, lettuce, sweet potatoes, and bananas — including the peel!

STEAM+Me

Bite into a carrot with your front teeth first. Now bite it with your back teeth. Which way was easier? Now chew a bite. Do you use your front teeth or your back teeth? Open wide and look at your back teeth in a mirror. Do you see how they have wide, flat surfaces that make them good for grinding and chewing? That's like a horse's teeth!

Different horses have different skills.

All horses have a lot in common: the same basic shape and abilities. But there are many **breeds**—or kinds—of horse, just like there are many breeds of dog. Each breed is a little bit different. People created some breeds of horse for doing special jobs. And some breeds were created just for their look.

Dancing Horses

Lipizzan stallions can walk in a graceful way that looks like marching or dancing.

Spot That Horse!

Appaloosas have bold, spotted markings. That makes them easy to tell apart from other horses.

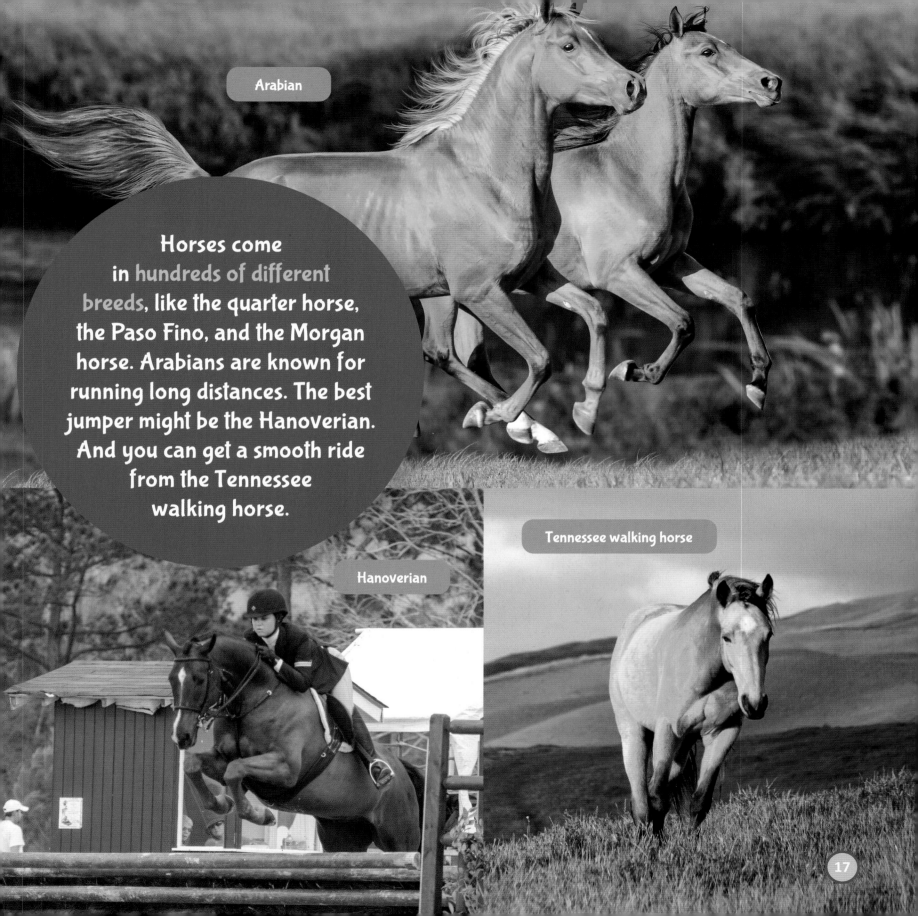

Arabian

Horses come in hundreds of different breeds, like the quarter horse, the Paso Fino, and the Morgan horse. Arabians are known for running long distances. The best jumper might be the Hanoverian. And you can get a smooth ride from the Tennessee walking horse.

Tennessee walking horse

Hanoverian

17

Let It Snow!

Dapple-gray horses almost look like they're covered in snowflakes.

Palomino

The Pink Horse

Horses with a rose-gray coloring can appear pink.

Every horse is beautiful.

Horses' faces are sometimes marked with splashes or flashes of white. Each kind of marking has a special name. The markings are called stars, blazes, snips, and strips. Horses' coats are different colors, and each color has its special name, too. Every horse has its own combination of markings and coat color that makes it unique.

Blaze

Star

STEAM☆Me

Pick two pictures of horses from this book to compare and contrast. Look at them carefully. What do they have in common? How are they different from one another? Look at their faces, markings, coloring, and size.

Meet the horse family.

Did you know there are other horselike animals in the world? Horselike animals are called **equines**. Other equines besides horses are donkeys, mules, and zebras. You can easily see why they are members of the horse family.

Horse

Equines all around the world have bodies and heads that are shaped similarly.

Mule

Mules Are Mixtures

A *mule* is an animal with a horse mother and a donkey father. An animal with a donkey mother and a horse father is called a *hinny*.

Donkey

Quagga!

The *quagga* was a kind of zebra. Quaggas had special coats and markings: the striped front of a quagga looked a lot like the front of a regular zebra. But the back part of a quagga was plain — it looked more like a horse. The last quagga died nearly 150 years ago.

Zebra

How do horses help us?

Horses are **domesticated** animals. That means they are tame and can live well around humans. People train domesticated horses to help with all kinds of jobs.

Four-Legged Post Office

Before cars and trucks, the Pony Express used riders and horses to deliver the mail. The Pony Express could deliver a letter the distance from Los Angeles to Chicago in 10 days. Today, delivering the letter by car would take about 30 hours.

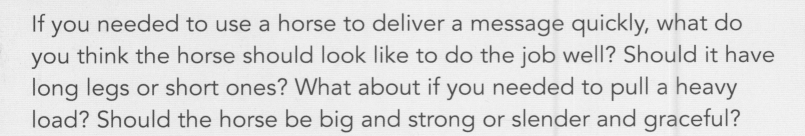

If you needed to use a horse to deliver a message quickly, what do you think the horse should look like to do the job well? Should it have long legs or short ones? What about if you needed to pull a heavy load? Should the horse be big and strong or slender and graceful?

STEAM #Me

Horses do jobs that involve pulling and carrying heavy loads. They also help people travel where there are no roads.

Ready, Set, Pull

Draft horses pull carts, wagons, and plows. These Clydesdale horses are a kind of draft horse. They have been used for carrying soldiers, plowing fields, and hauling logs. In a time before tractors, farmers could always turn to Clydesdales.

This is a job for a horse.

Horses work with people in many different ways. Horses help ranchers herd cattle. They help farmers haul supplies by wagon. They might also help police officers patrol a crowd. Horses can do lots of other jobs, too. Some of them might surprise you.

You might have heard of therapy dogs. They use their calm personalities to help people feel safe and comfortable. Therapy horses assist in a different way. They help people with disabilities develop strength and balance.

A Star Is Born

Horses can be movie stars! The horses you see in movies probably trained and practiced for a long time to be able to perform with human actors. Did you know that one of the first movies ever made starred a galloping horse?

Seek and Find

Horses are involved in search and rescue missions, helping to find lost and missing people. Their sensitive noses and ears help track down people in danger.

Where do horses come from?

Horses first appeared in North America millions of years ago. Then they made their way to Asia across a land bridge. Eventually, horses became **extinct** in North America but lived in many other places. Thousands of years later, explorers and travelers brought horses back to North America.

When horses first lived in North America, there were no humans around. Humans living in Asia started training and working with horses about 4,000 years ago.

If you could create a land bridge that would let you walk to any place on the planet, where would it take you? Draw the bridge and the place it leads to.

STEAM & Me

Welcome Home

When Europeans came to North America, they brought horses. Horses were once again living in North America.

Bridge Through Time

About 10 million years ago, North America and Asia were connected by a wide "bridge" of land. Horses used this to travel out of North America.

Some horses run wild.

Domesticated horses live with and around people. Other horses are wild. They live free, like other wild animals. When European people brought horses back to North America, some of those horses ran away. They started living by themselves, away from people. All wild horses alive today are related to the ones that ran away.

The wild horses of the western United States are known as *mustangs*. Mustangs live in herds with two leaders: One is a *mare*, which is an adult female horse. The other leader is a *stallion*, an adult male horse.

Wild and Rare

While there are millions of domesticated horses in the United States and Canada, there might be only about 60,000 wild horses left.

Living Free

Every summer, "saltwater cowboys" herd wild ponies from Assateague Island to Chincoteague Island, Virginia. It takes the horses about 10 minutes to make the swim.

What do you know about horses?

What things do you have in common with a horse? How are you different from a horse? Would you rather be a horse that can run really fast, a horse that can go very far, or a horse that is very strong? Imagine and draw a horse that you'd like to meet. Don't forget the markings on the horse's face and body.

Glossary

Learn these key words and make them your own!

breed: a kind of animal or plant that is related to others but different. *The Clydesdale is one of the strongest* breeds *of horse.*

domesticated: tame and able to work well with people. *It's easier to ride a* domesticated *horse than a wild horse.*

equine: an animal in the horse family. *Zebras are the only* equines *that have black-and-white stripes.*

extinct: no longer exists, often an animal or plant. *Dinosaurs went* extinct *millions of years ago.*

gait: a type of walking or running. *A horse's fastest* gait *is called a gallop.*

herbivore: an animal that eats only plants. *Horses eat only fruits, vegetables, grasses, and grains, making them* herbivores.

mammal: a warm-blooded animal that has a backbone, makes milk for its babies, and usually has hair on its body. *Humans, horses, and dogs are all types of* mammals.

ASP: Alamy Stock Photo; IS: iStock; LOC: Library of Congress; SCI: Science Source; SS: Shutterstock. Cover, Olga_i/SS; 5, Osetrik/SS; 6, Binnerstam/IS; 6-7, Alexia Khruscheva/SS ; 7, Wakila/IS; 8, Rapideye/IS; 9, Karri Huhtanen/Flickr; 9, (LO) Universal Images Group North America LLC/DeAgostini/ASP; 10, Lautaro Federico/SS; 10-11, Martin Nemec/SS; 11, Matylda Laurence/SS; 12-13, Kwadrat/SS; 13, (UP) everst/SS; 13, (LO) Alexia Khruscheva/SS; 14, Jaromir Chalabal/SS; 14-15, Marccophoto/IS; 15, Tagwaran/SS; 16, (CTR) Pegasene/SS; 16, (LO) olgaru79/SS; 17, (UP) Olga_i/SS; 17, (LO LE) S. Carter/Flickr; 17, (LO RT) Horse Crazy/SS; 18, (UP) Abramova Kseniya/SS; 18, (LO) Makarova Viktoria/SS; 18-19, Osetrik/SS; 19, (UP) nigel baker photography/SS; 19, (LO) Joss Chan/SS; 20, (UP) Svetlana Ryazantseva/SS; 20, (LO) Kdsphotos/Pixabay; 20, (LO LE) Sirisak_baokaew/SS; 21, (UP) mivod/SS; 21, (CTR) North Wind Picture Archives/ASP; 21, (LO) Jane Rix/SS; 22, LOC; 23, Robert Adrian Hillman/SS; 23, (LO) Mircea Costina/SS; 24, wrangel/IS; 24-25, Marcel Jancovic/SS; 25, ollo/IS; 26-27, LOC; 27, (UP) Andy Badenhorst/SS; 27, (CTR) Gary Hincks/SCI; 28, Mlenny/IS; 28-29, Greg Westbrook/SS; 29, smanter/IS; 30, (UP) Marcel Jancovic/SS; 30, (LO) Mircea Costina/SS; 31, (UP LE) Makarova Viktoria/SS; 31, (CTR LE) Kdsphotos/Pixabay; 31, (LO LE) olgaru79/SS; 31, (UP RT) Karri Huhtanen/Flickr; 31, (CTR RT) Tagwaran/SS; 31, (LO RT) Binnerstam/IS; 32, (CTR) Abramova Kseniya/SS; 32, (LO) JosephJacobs/IS; Back cover, (UP) ollo/IS, (LO LE) JosephJacobs/IS, (LO CTR) Abramova Kseniya/Shutterstock